Class 59s

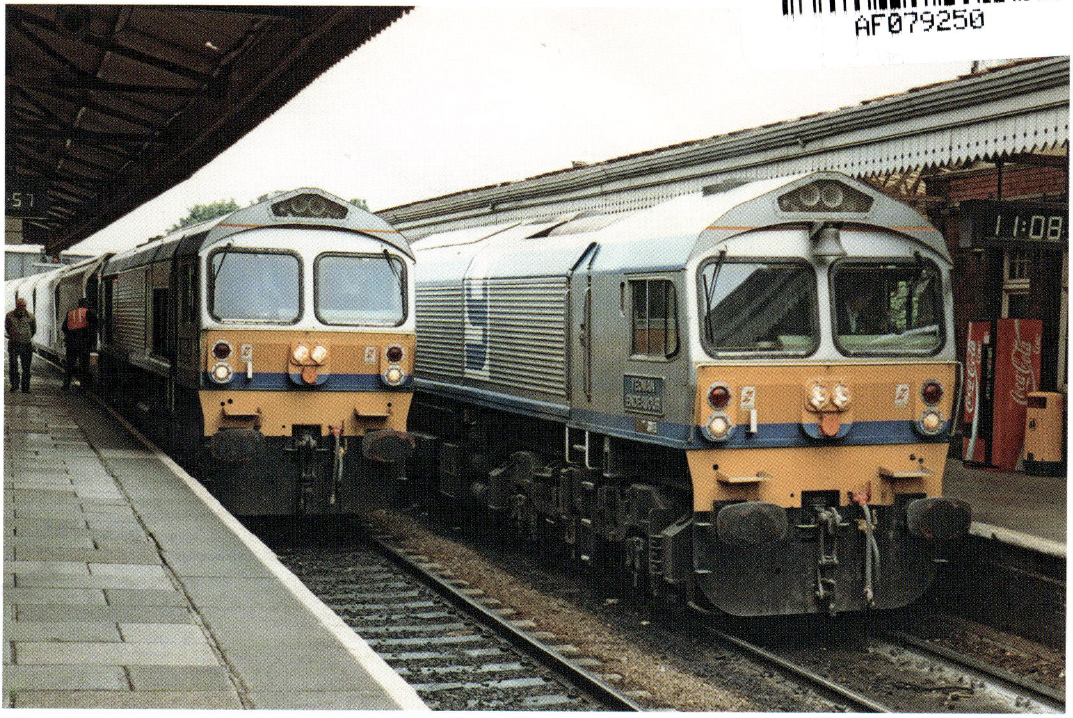

MARK V PIKE

KEY Books

BRITAIN'S RAILWAYS SERIES, VOLUME 25

Front cover image: Taken from the Warminster bypass bridge over the line near Upton Scudamore, 59003 is seen here having just reached the summit of the steep climb from Westbury with its usual 6O41, the 10.14 Westbury to Eastleigh East Yard engineers' train. 18 August 2020.

Back cover image: First of the class to receive the new Freightliner orange and black livery, 59206 is seen at Masters Foot Crossing near Fairwood Junction with 7A77, the 12.03 Merehead Quarry to Theale service. 26 August 2020.

Title page image: 59003 and 59001, both with empty trains, waiting to head west at Westbury station. Circa 1990.

Contents page image: 59104 at Westbury, next to a recently repainted 59103. 8 February 2006.

Published by Key Books
An imprint of Key Publishing Ltd
PO Box 100
Stamford
Lincs PE19 1XQ

www.keypublishing.com

The right of Mark V Pike to be identified as the author of this book has been asserted in accordance with the Copyright, Designs and Patents Act 1988 Sections 77 and 78.

Copyright © Mark V Pike, 2021

ISBN 978 1 802821 36 9

All rights reserved. Reproduction in whole or in part in any form whatsoever or by any means is strictly prohibited without the prior permission of the Publisher.

Typeset by SJmagic DESIGN SERVICES, India.

Contents

Introduction 4
Chapter 1   Class 59/0 5
Chapter 2   Class 59/1 49
Chapter 3   Class 59/2 71

Introduction

In the mid-1980s, Foster Yeoman, a major customer of British Rail for heavy block trains of stone, was becoming very concerned about the consistent reliability problems it was experiencing with the BR locomotives being used at the time, which were pretty much just Class 56s from around May 1983. Therefore, Foster Yeoman suggested to British Rail that its services could be better operated, and far more reliable, using a dedicated fleet of locomotives owned outright by the company. This was, at the time, a revolutionary idea and unheard of in the UK, no private owned traction had ever worked on BR before. However, BR recognised that Foster Yeoman was a very valuable customer and, eventually, an agreement was reached between the two companies, even gaining the support of the trade unions, for Foster Yeoman to obtain its own locomotives to be driven by BR drivers.

Foster Yeoman duly invited tenders for the supply of locomotives to the required specification. Although this tender was intended for any suitable builder, including in the UK, it finally went to the US firm General Motors, which had already provided Foster Yeoman with the powerful switcher loco (a kind of heavy shunter) used at Merehead Quarry. The deal was finally completed in November 1984 and the new locos were designated as Class 59 under TOPS.

As the locos were not a standard General Motors product of the time, they were constructed using their own dedicated production line with a specialised team of workers, and, amazingly, the four were completed in a little under six months. The locomotives arrived in the UK via Southampton Docks on 21 January 1986 and were unloaded over the following couple of days and then hauled to Westbury by a Class 47 loco. They were then commissioned and tested at Merehead and were later hauled to the Railway Technical Centre in Derby for further testing. As a result, both BR and General Motors declared the performance to be outstanding and measured a maximum tractive effort of 114,000lb. The class formally entered traffic on 17 February 1986.

The hauling power of these locos from the very start was excellent, with their availability being over 99 per cent, which actually exceeded Foster Yeoman's requirements. With the steadily growing business for Foster Yeoman still requiring regular hiring in of a BR loco, a fifth Class 59 was ordered in 1988, also being built as a special order and delivered in June 1989.

Other customers not happy with the BR diesels then being used for heavy block trains, such as ARC and National Power, were watching these developments closely. This resulted in ARC purchasing four Class 59/1s, which entered service during 1990, while National Power acquired six Class 59/2s during 1994–95. Both of these types now operate in conjunction with the original Class 59/0s hauling trains to and from the Somerset quarries of Whatley and Merehead, with occasional other work when required. The '59/0s' are now, incredibly, 35 years old, with the '59/1s' and '59/2s' close behind and still doing the work they were designed for.

Chapter 1
Class 59/0

59001 *Yeoman Endeavour*

Having entered revenue service just under two months before, this is one of my first images of the brand-new Class 59. It has just arrived at Westbury station with a Merehead Quarry-bound train and is undergoing a driver change. The initial four locos were only in service for about two months before they all received names, so this is during the short period when the locos were in service nameless. 4 April 1986.

Some six years later, the loco has now received its name and is seen at Westbury waiting to depart the station loop with another load for the London area. 8 April 1992.

This time we see the loco awaiting the signal at Reading station with an Acton to Merehead Quarry empty train. Notice the Network SouthEast-liveried Class 207 'Hampshire' unit in the bay platform awaiting departure with a stopper to Basingstoke. Reading station itself has undergone vast changes in recent years after a total rebuild. 18 October 1991.

This is Reading again, but this time heading in the opposite direction with a Merehead Quarry to Acton loaded train. Notice the different wagon styles that make up the load. At Acton, the train will be split into smaller sections and go forward to various terminals in the London area. 13 May 1992.

Class 59/0

Another view of the loco pulling away from Westbury heading to London. Notice the bell above the cab windscreens that was fitted upon completion of the loco in America. 18 October 1995.

After a change of livery, we now move to the South Western Main Line at Millbrook, near Southampton, with the loco heading west with the 7O51 Whatley Quarry to Hamworthy stone train. Occasionally, the bell receives a good polishing, but it had obviously not been done in a good while when this view was taken! 18 February 2005.

Seen from the station footbridge at Eastleigh, this is 7V16, the 12.47 Fareham to Whatley Quarry empty stone train. The sidings next to the station above the loco were little used at this time but, in recent years, they have become much busier as part of the virtual quarry that is now located here. 8 November 2007.

Producing a good amount of heat haze from its exhaust, the Whatley Quarry to Hamworthy stone train is seen again as it accelerates through Eastleigh station westbound. 3 April 2008.

The next change of livery came about in mid-2008 and has been in place ever since. This is a pristine loco, seen soon after receiving the latest Aggregates Industries (AI) colours at the little photographed location of Hamworthy Goods, near Poole, with the loco having run round its train and now ready to work the empty wagons back to Somerset. 17 September 2008.

The same train as seen in the previous image is now on the move as 7V52, the 13.58 Hamworthy to Whatley Quarry, and is captured crossing the calm water of Poole Harbour and heading along Holes Bay causeway, a few miles into its journey on the approach to Poole. 17 September 2008.

This time we see 59001 passing the site of the long-closed Patney and Chirton station, just west of Pewsey on the Berks & Hants line, with 6M20, the 10.37 Whatley Quarry to St Pancras Churchyard Sidings. Although the station has long since vanished, the footbridge that used to span the platforms, and from which this shot was taken, has been retained as a public footpath. 14 October 2009.

In 2016, the '59/0' sub-class reached the 30-year-old milestone and, to mark this, all of the Mendip pool Class 59/0s were adorned with this commemorative plaque. Although they were fashioned in the style of the cast name and number plates the locos have always carried, these plaques were actually stickers but, nevertheless, looked very convincing! 59203 is also seen in the background at Westbury. 5 October 2016.

Class 59s

Over the years, the class has often visited (mainly at weekends when they were not being used) various heritage lines in the UK. 59001 leans into the curve at Leigh Lane, near to Stogumber on the West Somerset Railway, with a Bishops Lydeard to Minehead service. This is a favourite line for the locos to visit, as it is not very far from their operating base. 11 June 2011.

Left: Still on the West Somerset Railway and on the same day, 59001 is seen at Tower Hill, near Williton, with a Minehead to Bishops Lydeard service. This time the bell is superbly polished. 11 June 2011.

Below: A final view on the West Somerset Railway sees the loco double-heading with 59103 *Village of Mells* and departing Blue Anchor with a Bishops Lydeard to Minehead train. 11 June 2011.

Class 59/0

This is the 6V18 Hither Green to Whatley Quarry empties, seen awaiting signals on the down through line at the old Reading station. This has always been an excellent place to see Class 59 activity. 17 February 2011.

A misty autumn morning at Winchfield, on the South Western Main Line, sees the loco passing through with 7O12, the 03.12 Merehead Quarry to Woking loaded train. 17 September 2014.

On one of the feeder services from the southeast that head for Acton, 59001 passes through Wandsworth Road in the London suburbs. On arrival at Acton, the train will join up with a couple of other services and form one 'jumbo' train that will head west to Somerset later in the day. 2 October 2014.

Just prior to the start of the Great Western Main Line electrification, this is the 6V18 Hither Green to Whatley Quarry empties approaching Twyford. As a result of the electrification, this view is now impossible, unfortunately. 4 March 2015.

Coming around the tight curve through Salisbury station is 7V09, the 11.41 Chichester to Merehead Quarry empties. This large station has hardly changed since it was built in the late 1800s. The train is passing the spot of the tragic train crash of 1906 that, unfortunately, killed 28 people when a London-bound express from Plymouth was derailed while running non-stop through the station at around 70mph when the line speed was only 30mph. This was reduced to 15mph as a result of this accident – a restriction that is still in place to the present day. 17 March 2015.

Unusually seen approaching Basingstoke on the up fast rather than the up slow line, 7O12, the 03.12 Merehead Quarry to Woking, makes a fine sight with a nice uniform rake of wagons. The background scene has changed here considerably since this image was taken, with the construction of a large high rise housing estate. 28 April 2015.

With the sunshine nicely illuminating the famous white horse on the hillside in the distance, this is the 6V18 Hither Green to Whatley Quarry empties arriving at Westbury station and passing the area signalling centre prominent behind the train. 27 July 2016.

Above: This time the uniformity of the train is spoiled by a couple of lower sided wagons as 7V16, the 11.47 Fareham to Westbury empties, makes its way through the Wylye Valley at Bapton, between Salisbury and Warminster. 26 April 2017.

Left: Passing a fine patch of Rosebay Willowherb in full flower at Heywood Village, just north of Westbury on the Trowbridge line, this is 7C29, the 06.24 Acton to Merehead Quarry. This train was diverted from its normal Berks & Hants route via Newbury at this time because of planned electrification work in that area. 10 July 2018.

The wheel flanges are squealing on the tight curve as 7V09, the 11.41 Chichester to Merehead Quarry empties, are eased past Redbridge Junction and on to the Romsey line. Many years ago, there was no fence here to impede the view, but in these days of health and safety awareness and potential trespass, it was considered necessary. 2 August 2018.

Nearing the end of its journey, this is 7A09, the 07.12 Merehead Quarry to Acton, approaching West Ealing. In the background are the buildings of the Plasser Works that deal mainly with on track machinery. Also brought into use in conjunction with the electrification here are the EMU sidings above the train, which were upgraded and now used for stabling GWR Class 387 units. 28 January 2019.

A popular location for photographers is Lambert's Bridge, just west of Westbury. This is heavy 6L21, the 13.23 Whatley Quarry to Dagenham Dock, passing the bridge and heading for the station, where there is usually a driver change before continuing its journey east. 14 May 2019.

Class 59s

There is a good blanket of snow at the pre-rebuilt Reading station as an unidentified train from Merehead Quarry bound for Acton passes through. 22 December 2009.

Having recently entered its 36th year of service, the veteran loco approaches Eastleigh with 6V62, the 13.21 Southampton Up Yard to Whatley Quarry. This train is a relatively recent addition to the Mendip Rail services, operating from the Somerset quarries. 5 July 2021.

Approaching Lambert's Bridge again, but this time from the opposite direction, 59001 had not long departed Westbury station with Pathfinder Tours' 'Fifty Niner' tour, 1Z58, the 06.40 Banbury to Merehead Quarry. 59002 was on the rear at this point. The Class 59s never seem to look quite right on a passenger train somehow! 20 February 2016.

59002 *Yeoman Enterprise / Alan J Day*

The loco is seen here pulling away from Westbury with a fully loaded train from Merehead Quarry bound for Acton. Note the Class 56 in the distance, the main loco type that Class 59s were built to replace. 16 March 1990.

Three years later, and heading in the opposite direction, the loco is seen arriving at Westbury with an afternoon train bound for Merehead Quarry. 26 March 1993.

With a nice trailing load of two-axle Yeoman PGA hoppers, this time the loco is heading west at Berkley Marsh, near Frome, with a Merehead-bound train. Thirty odd years on from the date of this picture, perhaps surprisingly, this view is still possible and perhaps even better after recent lineside clearance. Unfortunately, any view today would not include these wagons, as they were withdrawn many years ago. 9 February 1990.

By the time this image was taken, the two Mendip quarries of Foster Yeoman (Merehead) and ARC (Whatley) had joined forces to create Mendip Rail in 1993, and, as a result, the Foster Yeoman locos could regularly be seen hauling ARC wagons and vice versa. This loco is passing through the centre road at the original Reading station with the 6V18 Hither Green to Whatley Quarry service. That gas holder in the distance was quite a landmark for many years. 28 September 1995.

At this time just under a year old, 59002 is seen waiting at the signal at Lambert's Bridge, just outside Westbury, for permission to access Westbury Yard with a loaded northbound train. Once again, this view is still possible today, even after 34 years. 24 February 1987.

Another fine rake of Yeoman PGA hoppers from Merehead Quarry is viewed approaching Westbury. Lambert's Bridge, from which the previous image was taken, can be seen in the background.
14 September 1990.

By now, the loco had been renamed *Alan J Day* and is seen 12 years later at the same location as the previous picture with another load from Merehead Quarry. Note that 59002 is in the short-lived Mendip Rail green and orange livery that was only ever carried by this particular loco, which was an attempt to give the Class 59s a joint corporate identity, but it was not proceeded with.
28 June 2002.

The one-off livery can be seen to good effect in this image taken at the Old Oak Common open day. In hindsight, I personally think it was quite an attractive colour scheme, but, obviously, not everyone agreed!
5 August 2000.

59002 passes through Basingstoke with the 6V12 empty stone train from Woking to Westbury. 17 October 2001.

It is certainly not every day you will find a Class 59 on a steam age turntable, especially one that has remained in situ from that era! This loco, however, is on display on the Yeovil Junction table, one of very few original ones remaining (and fully working) in the country. The occasion was the first diesel day to be held at the Yeovil Railway Centre. 5 May 2003.

This time we see the loco heading west at Eastleigh with a rather short formed 7O48, the 09.23 Whatley Quarry to Hamworthy stone train. 27 August 2003.

Above: Slowly pulling away from Westbury station, this is another load from Merehead Quarry heading to the unloading stage at the Wootton Basset terminal, just west of Swindon. 4 February 2004.

Left: A view no longer possible owing to the removal of the bridge I was standing on, this is the empty 7V09 Chichester to Merehead Quarry train heading west past Millbrook. 8 April 2004.

With the low viewpoint accentuating the bulk of the train, another heavy load, the 7A17 Merehead Quarry to Acton, waits to depart the up reception line at Westbury. 8 September 2004.

Above: Heading down the 1 in 60 falling gradient southbound through Parkstone (Dorset) station in a summer shower, this is 7O48, the 09.23 Whatley to Hamworthy stone train, just a few miles from its destination. Note the rather ornate bridge in the background. 5 July 2005.

Right: By now carrying what was back then the standard Yeoman livery, this is 7V16, the 11.47 Fareham to Westbury empties, heading west through Salisbury station. 18 December 2008.

The well-known location of Fairwood Junction, near Westbury, sees the loco heading west with the lengthy 7C31 Theale to Merehead Quarry empties. The lines to the right form the station avoiding route. 25 September 2009.

Superb low autumn sunshine illuminates an Acton to Merehead Quarry empty service as it passes Berkley Marsh, near Frome, not long before sunset. 14 October 2009.

This is 7V07, the 12.41 Chichester to Merehead Quarry, seen on the Fareham to St Denys line, crossing the River Test near St Denys. 9 September 2010.

Perhaps surprisingly, after 35 years, it is still quite uncommon for any of the class to be used on anything apart from heavy stone trains to and from the Somerset quarries. However, when they were being used by EWS/DB Schenker, they did occasionally appear on infrastructure workings at weekends. This is the 6W41 Westbury to Blatchbridge Junction, Frome, via a loco run-round at Taunton at Wyke Champflower, near Castle Cary, conveying autoballaster wagons. 17 February 2013.

Now repainted into Aggregates Industries colours, the resplendent machine is seen finding a gap in the low winter shadows as it passes Farnborough with 7O12, the 03.12 Merehead Quarry to Woking service. 9 December 2015.

Captured at Sherrington public foot crossing in the Wylye Valley, between Salisbury and Warminster, this is the 7V07 Chichester to Merehead Quarry empties. 10 July 2018.

Even after 30-odd years, the locos always look very clean and smart, a testament to all the staff at Mendip Rail. This is the heavy 6L21, the 13.23 Whatley Quarry to Dagenham Dock loaded train, at Berkley Marsh, near Frome. 28 March 2019.

Coming through the down reception line at Westbury station just prior to sunset, this is the 6V18 Allington to Whatley Quarry empties. This view has now become available since the platform extension to accommodate ten-car Class 800/802 IET formations was completed a couple of years ago. 4 November 2020.

59003 *Yeoman Highlander*

Reading station again, with another loaded train from Merehead Quarry heading for Acton passing through the original platform nine. The centre line was often used for running round locomotives when arriving on inter-regional passenger services to and from the south to the north of England. Circa 1992.

As mentioned earlier in the book, after the merging of the rail operations of ARC and Foster Yeoman in 1993, the '59/0s' could be seen hauling ARC wagons around the network and initially looked a bit odd. This is the 6V18 Hither Green to Whatley Quarry passing through the down reception line at Westbury station. Since this picture was taken, the platform that is just in view to the left has been extended, rendering this view no longer possible. 6 March 1996.

Three fifths of the Class 59/0 fleet are seen at Westbury in one image, with 59003 in the distance arriving from Wootton Bassett, 59001 in the centre arriving from Acton and 59005 to the right waiting at signals to proceed light engine. Circa 1990.

This particular loco has had a rather more eventful life than the rest of the class. During 1997, it was exported to work (albeit stored unserviceable for various periods of time) in Germany and continued to do so on and off until its withdrawal from service there in 2014. It was then that GB Railfreight purchased it, and the loco was repatriated in 2015 for refurbishment and return to UK standards at Eastleigh Works. By mid-2015, it was ready to go on test and is seen approaching Fareham with its first run under its own power on the UK network since the end of 1996, along with 66717 *Good Old Boy* for insurance, just in case of any problems. It will be noted that, at this point, the paintwork was not quite finished around the bufferbeam skirting. After this run it was deemed good to go. 29 May 2015.

Some of 59003's first solo workings took place on the West Somerset Railway at the diesel gala held there in 2015. As well as being quite a draw for visitors to the gala, it also helped to run the loco in and iron out any teething troubles that might be encountered. It is seen between Stogumber and Williton with a Bishops Lydeard to Minehead service. 5 June 2015.

Just like the old days! Once in revenue service, it was not long before it was again working stone trains off Westbury. The rather dusty looking loco is seen departing Westbury station with 6M40, the 11.55 Westbury to Cliffe Hill Stud Farm empties. 15 January 2019.

Beneath a mass of electrification equipment, this particular train, the 7B12 Merehead Quarry to Wootton Basset, is approaching Swindon and heading for the loop to the east of the station to enable the loco to run-round, as there is no access to the terminal from the westerly direction. This was the first train the loco had worked from Merehead Quarry since late 1996. It was during a very short time, just a matter of days, when the loco was on hire to Freightliner, which had just secured the Mendip contract. 15 November 2019.

After a few trials on various GBRf operated trains in different parts of the UK, 59003 eventually settled down on 6O41, the 10.14 Westbury to Eastleigh East Yard engineers' train, a regular Mondays to Fridays service that it still has a monopoly on at the time of writing. The train is seen passing Salisbury station, on this day conveying an empty long welded rail set. 11 December 2019.

The return working of the train mentioned in the previous picture is 7V41, the 14.45 Eastleigh East Yard to Westbury, seen about to pass over Mount Pleasant level crossing, just south of St Denys, on a fine winter's afternoon. 20 January 2020.

7V41 is viewed again, this time passing Salisbury station, with the loco now sporting a newly revised livery style; the orange cabsides have been extended back to now include the cab doors and remove the obsolete Europorte markings. Full marks to GBRf for retaining the original cast name and number plates, along with the data panels in keeping with the rest of the sub-class. 20 July 2020.

Taken from the Warminster bypass bridge over the line near Upton Scudamore, 59003 had just reached the summit of the steep climb from Westbury with its usual 6O41, the 10.14 Westbury to Eastleigh East Yard engineers' train. 18 August 2020.

Another fine spot to capture 6O41 is at Little Langford in the Wylye Valley, between Warminster and Salisbury. On this day, the train was made up of a real motley collection of wagons behind the loco. 12 April 2019.

Another good load for 6O41, this time approaching its terminating point as it slowly creeps up to the signal at Campbell Road bridge, Eastleigh. 9 December 2020.

Another heritage line to have seen the loco used on passenger services was the Swanage Railway, during its 2019 diesel gala. It did, however, have to work with a translator loco to operate the train's vacuum brakes. 59003 is seen piloting 73119 *Borough of Eastleigh* soon after departing Harman's Cross bound for Norden on a fine spring morning. 11 May 2019.

This time 59003 is captured at Harman's Cross station, again in the company of 73119 *Borough of Eastleigh*, heading a Swanage to Norden service. It is very hard to believe that until the Swanage Railway preservationists took over this delightful branchline in the mid-1970s after closure by BR, a station never even existed here. 11 May 2019.

59004 *Yeoman Challenger* / Paul A Hammond

Virtually brand new, the loco is seen stabled at Westbury. Various locos can still be found laying over at this point today, but the building with the vents on the roof has long since gone and, as mentioned earlier, the platform has been extended in the foreground at this point. 4 April 1986.

Another view of a wonderful rake of the old Yeoman PGA hoppers, the train is approaching Lambert's Bridge on the approach to Westbury with a Merehead Quarry to Botley loaded working. 1 September 1989.

The same train seen in the previous picture has now changed direction in Westbury Yard, with the loco having run round. It is seen climbing the steep incline up past the small station of Dilton Marsh, at not much more than walking pace, on its way east via Salisbury to Eastleigh and Botley. 1 September 1989.

A view that is now lost from the footbridge near West Drayton sees 7C77, the 12.41 Acton to Merehead Quarry, approaching on the down local line. Not only has electrification done in this view, but the footbridge has been replaced by one of those awful, caged monstrosities. By this time, the loco had received its *Paul A Hammond* nameplates. 5 December 2008.

Yet another location that has changed completely, with both electrification and the total removal of the footbridge I was standing on. Passengers now have to exit/enter the station platforms by means of a ramp down to the main road, then under the bridge the third wagon is pictured passing over. 7C77, the 12.41 Acton to Merehead Quarry, is seen again as it heads through Reading West station. Reading Depot in the background has since been relocated out of sight beside the main line to Bristol/South Wales, but sidings are retained on the old depot site for stabling and servicing track machines and Network Rail test trains, now known as Triangle Sidings. 21 January 2009.

We have now moved up to the very eastern end of the Great Western Main Line where 7A09, the 07.12 Merehead Quarry to Acton, is passing West Ealing near the end of its journey. Even this station has been modernised in recent years, most noticeably by the extension of the platform I am standing on almost as far as the set of points in the distance and the construction of a bay platform behind the station nameboard for the use of trains to/from the Greenford line. 6 November 2009.

Looking a little lost, this is 59004 running light engine towards Westbury Yard past a familiar spring time sight of a bright yellow rapeseed field just west of the station. It is not that common to capture a Class 59 running light engine. 21 April 2011.

Once a very familiar landmark in the Wiltshire town of Westbury, the 400ft (121.9m) tall chimney, which used to form part of the cement works, was demolished in September 2016 watched by a large number of the town's people. Here we see 7C77, the 12.41 Acton to Merehead, again passing by the huge structure and taking the line towards the station. 13 January 2012.

Powering through Kensington Olympia, this is one of the various feeder services that converge on Acton during the day to form up one or two longer trains that will head west later in the afternoon back to one of the Somerset quarries. 6 November 2003.

Unusually looking a little work-stained and weary, the loco dodges the morning shadows just east of Basingstoke, powering 7O12, the 03.12 Merehead Quarry to Woking loaded train. Notice also that, by this time, the Foster Yeoman branding amidships had been removed. 26 February 2014.

Slowly creeping to a halt at signals at Westbury station, this time the loco has the company of 66168 on 7A09, the 07.12 Merehead Quarry to Acton service. Double-heading with Class 59s, or any other class of loco, is not something done to provide extra power, but usually just to move locos around the system to avoid any extra light engine moves. 10 April 2014.

In the mid-2010s, the loco became something of a minor celebrity when it was the final example still in traffic in the obsolete Yeoman livery. It is seen passing through Salisbury, shortly before it was admitted to the paintshop with the 7V07 Chichester to Merehead Quarry empties. 9 April 2015.

In the now standard Aggregates Industries livery, the loco is again seen at Salisbury, this time approaching from the west with 7O12, the 03.12 Merehead Quarry to Woking service. The South Western Railway carriage washing facilities can be seen behind the train; these facilitates form part of Salisbury Depot, which maintains the near 30 year old fleet of Class 158/159 diesel units used on West of England services.
2 March 2017.

Coming slowly under Lambert's Bridge, this is 7A09, the 07.12 Merehead Quarry to Acton. The old wooden fence posts in the foreground are a relic of the original Great Western Railway and somewhat enhance the image.
5 April 2017.

With the newly installed overhead catenary now prominent, this is 7A09, the 07.12 Merehead Quarry to Acton once again, this time drifting through Twyford, a few miles east of Reading.
1 September 2017.

Powering off the Salisbury line at Worting Junction, a couple of miles west of Basingstoke, this is 7O12, the 03.12 Merehead Quarry to Woking again. This is the point where the line to Salisbury and Exeter leaves the Waterloo to Weymouth main line. 10 August 2016.

Presenting a powerful image, this is 7C29, the 06.41 Acton to Merehead empties, approaching Masters public foot crossing just west of Fairwood Junction, which is located near to the bridge just visible in the distance. 26 August 2020.

Organised by Pathfinder Tours, back in the late 1990s and early 2000s, Mendip Rail used to run an annual charter train for its staff and their families to various parts of the country, which often took members of the class into unusual territory. Having both received a bit of a spruce up, this is 59004 in the company of 59103 *Village of Mells* arriving at Castle Cary to pick up passengers with 1Z59, the 08.08 Bristol Temple Meads to Par (for the Eden Project), which the locos worked throughout. All three sub-classes have never been common in the Duchy of Cornwall. The headboard carried Union flags as part of the Queen Elizabeth II special jubilee celebrations being held that year. 3 June 2002.

Another loco balancing move, this time utilising 7C77, the 12.41 Acton to Merehead, with 59004 leading 59201 into Westbury. Note the white horse on the hillside in the distance and the GBRf Class 66/7 stabled behind the train. 27 July 2016.

59005 *Kenneth J Painter*

This loco is a little newer than the rest of this sub-class, having entered service in 1989, some three years after the original quartet. This was due to the steady increases in traffic identified by Foster Yeoman at the time and the need for another loco to cover this, rather than continuing to hire in a BR loco. On the night of 25–26 May 1991, a haulage trial involving this loco to test the operation and feasibility of longer trains from Merehead was held on the main line near Frome. The train was made up of a remarkable 115 wagons, weighing 11,982 tonnes and 5,415ft (1,650 metres) long, which was then the heaviest and longest train ever operated in Europe. To mark this achievement, the loco received a commemorative plaque on the cabside (just beside the fleet number in this image) that it still carries today. Just a couple of years old at the time, it is seen pausing for a signal check at the old Reading station with an Acton to Merehead Quarry empty train. 3 December 1991.

This image sees the loco heading westbound through Kensington Olympia with a feeder service from Acton to one of the terminals in the southeast. 29 August 1992.

Class 59/0

An unidentified westbound empty train is seen waiting to depart platform one at a rather damp Salisbury. In recent years, this platform has been used as a reception line only for Salisbury Depot and not used by passenger trains. There have been various rumblings about restoring it to full use but, as yet, nothing has come of them. 22 March 1991.

Another one of the rare occasions during the EWS/DB Schenker jurisdiction (especially on a weekday) when one of the class was used on an ordinary freight train, as opposed to the bulk trains working to/from the Somerset quarries. This is 6O41, the 10.14 Westbury to Eastleigh East Yard, passing Millbrook container terminal. One can only assume that there was a shortage of the usual locos at Westbury that morning. Since the takeover of the Mendip contract in late 2019, Freightliner has yet to use any of the class away from working block aggregates trains. 25 January 2006.

The classic location of Campbell Road bridge at Eastleigh sees the loco heading south with the 7O48 Whatley Quarry to Hamworthy loaded train. 3 August 2005.

In the dying rays of the autumn sunshine, this is the 6V18 Hither Green to Whatley Quarry arriving at Westbury. 7 November 2014.

Above: This time we see 59005 approaching West Drayton, making for a rare sight hauling Freightliner wagons (on hire to EWS at the time) on a loaded but unidentified working that only went to West Drayton sidings to run round, returning east soon after (see next image). It is rather ironic that around 11 years after this image was taken, the whole Mendip contract, including this train and loco, actually went over to Freightliner operation! 5 December 2008.

Right: Half an hour later, the same train now comes back through West Drayton heading back east, presumably to Acton. It is unclear what the reason was behind this unusual move. 5 December 2008.

With a two-car Class 165 'Turbo' unit on a Reading to Basingstoke service heading in the opposite direction, the 7A17 Merehead Quarry to Acton passes Southcote Junction, near Reading. 21 January 2009.

The very popular spot for photographers at Fairwood Junction, just west of Westbury, is the location again for this image of 6M20, the 10.38 Whatley Quarry to St Pancras Churchyard Sidings. Many years ago, there used to be a delightful old wooden signal box behind the first and second wagons of the train and some huge Elm trees behind that. These trees were killed off during the Dutch Elm disease of the early 1970s but now finally appear to be slowly returning. However, the signal box will definitely not be returning.
25 September 2009.

Seen from the public footbridge just to the east of the station, 59005 passes through the centre road at Redhill with 6V60, the 09.26 Ardingly to Acton. 1 October 2009.

This is 7V52, the 13.58 Hamworthy to Whatley Quarry empty stone train, approaching Southampton Central.
12 May 2011.

Slowly approaching an adverse signal at Salisbury platform four, this is 7O12, the 03.12 Merehead Quarry to Woking. 18 September 2014.

Rounding the long curve by the picturesque Kennet and Avon Canal at Crofton, 59005 heads west with the 6V18 Allington to Whatley Quarry empties. The area around Crofton has long been a favourite location for generations of railway photographers. Having the canal very close by is, of course, an added interest. 8 April 2015.

This is 7C29, the 06.41 Acton to Merehead Quarry empties, soon after departing Westbury and approaching Lambert's Bridge. The famous white horse can just be seen in the mist on the hillside in the distance. 5 April 2017.

The harvest has just begun in these parts as we see a nice uniform rake of empties that form 7V16, the 11.47 Fareham to Westbury, rolling past the hamlet of Bapton, in the Wylye Valley between Salisbury and Warminster. 18 July 2017.

Chapter 2
Class 59/1

59101 *Village of Whatley*

The second private company to secure a small fleet of four locos was ARC, and its Class 59/1 locos entered service during late 1990. They were of a revised design from the original locos with the most obvious difference being the front-end light clusters, which were of the now familiar style used on UK railways. In the original mustard coloured ARC livery, the loco heads through Reading with an Acton-bound loaded train. 6 April 1994.

Still looking quite new after just over a year in service, the loco scatters the local pigeons at Eastleigh as it leads a Fareham to Whatley Quarry empty train through platform two, where it will soon cross the South Western Main Line and head onto what was then the freight-only line to Romsey. 11 February 1992.

Another instance of mixed up loco and wagons as a rake of Yeoman empties arrives at Westbury heading for Merehead Quarry. 2 November 1994.

A close-up of the loco in its original ARC livery waiting at Westbury with a loaded eastbound train. 24 February 1993.

Now in the very familiar Hanson livery that all the Class 59/1s have worn for over 20 years, this is a diverted 7A09, the 07.12 Merehead Quarry to Acton, approaching Didcot Parkway. The reason for the diversion was signalling problems on the usual, and more direct, Berks & Hants route via Newbury. 10 March 2011.

An almost perfect reflection of 59101 as it runs eastwards alongside the Kennett and Avon Canal near to Crofton with 7A09, the 07.12 Merehead Quarry to Acton. 30 September 2011.

The loco is seen in Acton Yard soon after arrival with another heavy load of stone. The train will no doubt be split here and depart for various unloading locations in and around the southeast. 4 March 2015.

Coming round the tight curve off the main West of England line on the approach to Westbury station, this is a Theale to Merehead Quarry empty train. Many trains were originally routed via the station rather than the avoiding line for crew change and pathing purposes. However, in recent years, many now use the avoider, especially under the Freightliner reign, as the yards at Westbury continue to be DB-operated and there are charges to be paid when using them! 29 March 2004.

At first glance, this might appear to be one train with a loco on each end, but it is in fact two trains passing at Heywood Junction. 59101 is heading east with a loaded train, while passing in the opposite direction with a train of empties is 59204. The image was taken from high up on White Horse Hill, which is a super viewpoint with commanding views across the whole area. However, I suspect that the lineside itself is not nearly so visible now, with the relentless undergrowth having taken over again. 14 October 2009.

In complete contrast to the last image, 59101 is seen at Ealing Broadway in the western suburbs of London with 7A09, the 07.12 Merehead Quarry to Acton loaded train. Note the London Underground unit glimpsed to the right. 24 June 2015.

Powering through Basingstoke station, this is 7O12, the 03.12 Merehead Quarry to Woking service. 18 April 2018.

Another location scuppered by the overhead electrification, this is good old 7A09, the 07.12 Merehead Quarry to Acton again, this time passing through Slough on the last few miles of its journey to its terminating point. 15 April 2014.

Rounding the long curve at Crofton on the Berks & Hants line, the loco has ex works 59202 for company on 7A09, the 07.12 Merehead Quarry to Acton. Since the date this picture was taken, the area around here has had a bit of a makeover, with the trees and bushes trimmed back. Pity it will not stay like it for long! 30 August 2012.

Amazingly, because of such a lengthy signal check the train encountered here just after the last picture was taken, I was able to get to this foot crossing just to the east to obtain another shot of the train just getting into its stride again. The old pumping house and chimney can be seen through the heat haze of the exhaust. 30 August 2012.

59102 *Village of Chantry*

Looking quite smart with a rake of uniform liveried ARC wagons, this is the 6V18 Hither Green to Whatley Quarry empties passing Reading. Just about the only part of this station that survived the major rebuilding of the mid-2010s was the listed part with the clock tower, seen behind the loco. 18 June 1992.

Once again hauling a nice uniform rake of wagons, the loco is seen awaiting the signal at Salisbury station to proceed with an empty train bound for Whatley. Just in view behind the wall to the left of the loco is the Salisbury Maintenance Depot where the SWR Class 158/159 units are maintained, and which had recently opened when this image was taken. The depot is actually built on the site of the former GWR station that used to exist here. 9 October 1993.

My first view of one of these new ARC locos. 59102 is just four months old here and yet to receive its name as it brings a loaded train slowly through Salisbury station heading for Eastleigh. Older style wagons than the previous picture this time but still in the matching ARC livery. The platform from which this image was taken has since been taken out of passenger use. 22 March 1991.

This is the 6V18 Hither Green to Whatley Quarry empties again, this time arriving at Westbury. 18 October 1995.

With a nice bit of trackside clearance having recently taken place along this stretch of line, this is the 7O51 Whatley Quarry to Hamworthy stone train approaching Hinton Admiral on the edge of the New Forest National Park, and on the main Waterloo to Weymouth line. Since this picture was taken, the trackside has once again reached jungle proportions! 28 November 2007.

59102 is seen passing the rarely photographed location of Stert, in an equally rare gap in the lineside vegetation, between Woodborough and Lavington on the Berks & Hants line, with an unidentified eastbound loaded train. 14 October 2009.

Right: Prior to the Great Western Main Line electrification, the old GWR-built footbridge at Taplow station was a great platform from which to obtain shots such as this with a telephoto lens. Here we see 7A09, the 07.12 Merehead Quarry to Acton, on the approach to the station. 5 January 2012.

Below: Laying an impressive, though not if you happen to have your washing out, dust screen across the countryside, this is heavy 6L21, the 13.23 Whatley Quarry to Dagenham Dock loaded train approaching Little Bedwyn on the Berks & Hants line. 30 August 2012.

Heading south through Oxford station, this is the Oxford Banbury Road to Whatley Quarry empties. As is the case at many of the larger stations on the network, the car parking area to the right was a former goods yard. 5 December 2014.

Apart from the annoying graffiti on the second wagon, the whole train looks very smart in the crisp winter afternoon sunshine. This is empty 6A83, the 11.35 Avonmouth Bennetts Road to Westbury service, passing through the Avon Valley at Claverton, just south of Bathampton. This is a train that now only occasionally runs. 14 February 2015.

More lineside clearance at the time opened up this long-lost shot of diverted 7C29, the 06.24 Acton to Merehead Quarry empty train, passing Hawkeridge Junction near Westbury. The train was diverted to run via Swindon, Melksham and Trowbridge because of electrification work in the Newbury area. The lines to the right of the picture lead to the main line from the West of England to Paddington at Heywood Road Junction. 26 April 2018.

A chilly winter's morning at Hook on the South Western Main Line sees 7O12, the 03.12 Merehead Quarry to Woking, passing through the early morning shadows. 17 February 2015.

Seen in an always popular photography spot, this is 7C31, the 08.40 Theale to Merehead Quarry empty stone train, passing by the canal and lock gates at Crofton on the Berks & Hants line. Note the ex-EWS coal hoppers, which had been redeployed to convey stone, that were in use at this time. 8 April 2015.

This is 7A09, the 07.12 Merehead Quarry to Acton, approaching Taplow station on the up local line; the main lines are to the left of the picture. This is the only reasonable shot now obtainable here following electrification, taken from the platform end. 20 May 2019.

Class 59s

59103 *Village of Mells*

Curving its way round Eastleigh station, this is a Whatley Quarry to Eastleigh loaded train. Note that the rear of the train is still coming off the Romsey line in the left distance. 15 July 1994.

Then only just over a year old, and as yet unnamed, the loco passes through Salisbury with an eastbound loaded train. The BRSA (British Rail Staff Association) club in the background has since been demolished. 21 May 1991.

Class 59/1

After the loco had run round in the yard, this is 7V16, the 11.47 Fareham to Whatley Quarry empties, coming round the curve from Westbury station. The white horse is just visible on the hillside in the distance. 26 April 2018.

Many miles distant from the previous image, 59103 passes through Denmark Hill in the leafy London suburbs, not long into its journey with the 6V18 Hither Green to Whatley Quarry empties. 19 May 2011.

This time we see the loco in the heart of the New Forest as it approaches Lymington Junction, just west of Brockenhurst, with the 7O51 Whatley Quarry to Hamworthy stone train. The line to the right is the Lymington branch. Not technically a junction anymore, the line to Lymington has been a separate track from Brockenhurst station (out of sight past the rear of the train) since the mid-1970s. Originally, there was a signal box and a three-way junction on the other side of the bridge from which this picture was taken. 2 August 2006.

Once again seen a long way from home, this is an empty train crossing the attractive viaduct at Eynsford, between Sevenoaks and Swanley, bound for Acton and eventually one of the Somerset quarries. 29 September 2011.

Class 59/1

This image is again taken at Crofton but this time heading in the opposite direction as we see 7A09, the 07.12 Merehead Quarry to Acton service. The pump house that is home to the magnificent steam-powered beam engines can be seen behind the train.
30 August 2012.

A little further east from the previous image, we see an Acton to Whatley empty train passing the attractive location of Little Bedwyn. The Kennett and Avon Canal seen to the right fell into decline during the 1950s, but after some successful campaigning, the 87 miles were renovated in sections and finally it was fully restored and reopened in 1990. It is now very popular with holidaymakers and locals alike.
30 August 2012.

Back on the South Western Main Line in Hampshire, this is 7V52, the 13.58 Hamworthy to Whatley Quarry empty train, approaching Southampton Central beneath a threatening autumn sky.
22 October 2010.

Much nearer to home now as we see 59103 at Heywood Village, soon after leaving Westbury with re-timed and re-routed 7A09, the 06.27 Merehead Quarry to Acton, which was diverted because of electrification work in the Newbury area on its booked route. 10 July 2018.

As we have seen with the Class 59/0s, some representatives of the Class 59/1s have also been used on heritage lines over the years, and these three images show 59103 at work, once again on the West Somerset Railway. This image shows the loco paired up with GBRf's 66716 and rounding the curve at Roebuck, just south of Crowcombe Heathfield, with a Bishops Lydeard to Minehead train. The 'Bluebird' was quite new at the time, and in fact this was the first time a Class 59/1 and a Class 66/7 combination had occurred. 9 May 2004.

Later the same day, the loco returned to Bishops Lydeard with EWS-liveried Class 37/4 37419 for company. The pair are seen approaching Leigh Woods level crossing between Stogumber and Crowcombe Heathfield. 9 May 2004.

Seven years later, the same loco was again used on the West Somerset Railway, this time in the company of stable companion 59001. The pair are seen arriving at Crowcombe Heathfield with a Minehead to Bishops Lydeard train. Notice the temporary huge American-style warning horns on the front of the loco! 11 June 2011.

59104 *Village of Great Elm*

Arriving at the familiar location of the up reception line adjacent to the station at Westbury, this is a loaded train bound for Acton. The car auction building in the background is still in use today. 8 April 1992.

Leaning to the curve on the down through line at the original Reading station, this is 7C77, the 12.41 Acton to Merehead Quarry empties. Note the once familiar landmark of the gas holder in the background, which has featured in generations of railway photographs through the years. 3 August 1995.

Class 59/1

Slowly passing through platform two at Eastleigh is a Whatley Quarry to Botley loaded service. Although Botley is still served by stone trains from the Mendips today, the only regular train that serves it now runs overnight. 11 February 1992.

Heading across the causeway at Holes Bay, part of the large expanse of Poole Harbour, this is the 7O51 Whatley Quarry to Hamworthy loaded train about a mile or so from its destination. 18 October 2004.

Considering the Class 59/1s have carried the Hanson livery for over 20 years now, it has worn very well. This is a profile of 59104 at Westbury. 14 September 2020.

A pleasant spring morning sees the 7B12 Merehead Quarry to Wootton Basset loaded train passing Hawkeridge Junction, soon after departing Westbury. The lines leading off to the left of the picture are the station avoiding lines and are often used for passenger train diversions when engineering work is taking place on the main Swindon–Didcot route. 5 April 2017.

At the time, this was one of the few locations in the lovely New Forest National Park that was clear enough to obtain a reasonable shot of trains. This is the 7O51 Whatley Quarry to Hamworthy loaded stone train approaching Beaulieu Road. Inevitably, it has now become very overgrown here at the time of writing. 25 March 2011.

After 30-odd years in service, it is still surprisingly uncommon to observe a pair of Class 59/1s double-heading and, with this short-formed train, it was certainly not for any extra power! The 6V18 Hither Green to Whatley Quarry is captured approaching West Drayton with 59104 having 59103 for company. It also helps the overall appearance when the wagons are in the same livery as the locos, as is the case here. 31 August 2012.

On a superb autumn morning, this is 7O12, the 03.12 Merehead Quarry to Woking loaded train, rattles across the pointwork through Basingstoke. 28 September 2016.

The usual view one sees of the Westbury white horse is from afar, but it is a little larger when you are right next to it! This is the super view across Wiltshire afforded from the top of White Horse Hill. Looking more like a model, 59104 passes the old cement works with 7C77, the 12.40 Acton to Merehead Quarry empties. As an added bonus, an RAF Hercules transport plane passes over at just the right time! 14 October 2009.

Another double-header as 7A09, the 07.12 Merehead Quarry to Acton, is seen yet again, this time as 59104 heads through West Drayton along with 59201 *Vale of York* as company. 15 September 2011.

Chapter 3
Class 59/2

59201 *Vale of York*

Above: Powering through Eastleigh on a fine winter's day, this is the 7O51 Whatley Quarry to Hamworthy loaded stone train. Much of the track and pointwork had been recently renewed at this date, hence the clean looking ballast. 12 January 2005.

Right: With the loco looking clean and smart, this is returning 7V52, the 13.58 Hamworthy to Whatley Quarry empties, a few miles into its journey approaching Poole station. Poole stadium can be seen in the background. 12 September 2007.

59201 passes through Taplow station with 7C77, the 12.41 Acton to Merehead Quarry empties. Like most of the formerly good photographic locations along the Great Western Main Line, this shot is now impossible owing to the overhead catenary and the small matter of no longer being able to access the platform on which I was standing. 19 November 2008.

A study of the loco stabled next to Westbury station, clearly showing the commemorative bell attached to the cab front on this end only. When clean, the maroon EWS livery did look quite smart. Curiously, while 59001 and 59201 received this embellishment, 59101 did not. 28 February 2005.

59201 approaches Eastleigh with one of its first workings after being painted in DB Schenker livery, although now having lost its nameplates. The immaculate loco is at the head of 6O41, the 10.14 Westbury to Eastleigh East Yard engineers' train. The brass bell really stands out well after a good polish. 2 May 2012.

Heading west through the lush green countryside at Lavington on the Berks & Hants line, this is the 6V18 Allington to Whatley Quarry empties. 30 August 2016.

This is the unusual sight of the loco at the head of the Belmond British Pullman train. This was one of the rare occasions the Pullman stock had been hired by a private tour operator and was running as 1Z59, the 08.15 London Victoria to Minehead (West Somerset Railway) 'Quantock Pullman', organised by UK Railtours. 59201 is seen soon after departing Westbury and heading west. Classmate 59205 was out of sight on the rear of the train at this point. 14 May 2016.

Being a dedicated freight loco, all three sub-classes can often be requested for railtour duty. This was the case here as we see the loco passing Bradford on Avon (with 59206 on the rear) with UK Railtours' 'The Somerset Strimmer', which visited various unusual freight only and secondary lines in and around Somerset. 16 March 2019.

A further view at Bradford on Avon, but now heading in the opposite direction, this is the same charter as in the previous picture returning west after visiting Avonmouth. The local name for this spot is 'The Avenue', which refers to the rows of large Cedar trees along the lineside that the train has just passed through. This was a favourite spot for the great Somerset & Dorset Railway photographer Ivo Peters back in the 1960s and 1970s, who lived nearby in Bath. 16 March 2019.

Back to more usual duties now. Rounding the curve past Hawkeridge Junction, which is situated just past the bridge in the background, this is 7B12, the 11.34 Merehead Quarry to Wootton Basset loaded train, which has just departed from Westbury station. 15 August 2017.

Across to the southeast now in Kent as we see the 6V18 Allington to Whatley Quarry empties approaching Tonbridge station. The train is coming in from the Paddock Wood direction, while the tracks to right of the picture head off towards Tunbridge Wells.
19 October 2016.

Right: Back during its time working in the early days of DB Schenker, the loco is seen with 59104 *Village of Great Elm* for company approaching West Drayton with 7C77, the 12.41 Acton to Merehead Quarry empties.
15 September 2011.

Below: Double-heading again, this time with 66063 on 7A09, the 07.12 Merehead Quarry to Acton at Ruscombe, just east of Twyford on the Great Western Main Line.
1 October 2015.

59202 *Vale of White Horse / Alan Meadows Taylor*

For a short while in the mid-2000s, the '59/2' sub-class was trialled by EWS on the Avonmouth to Furzebrook (Dorset) liquefied petroleum gas trains, which have long since vanished from the system. These were usually Class 60 or Class 66 hauled at the time. The loco is seen exiting Southampton Tunnel with 6W53, the 08.45 Eastleigh Yard to Furzebrook. 23 December 2004.

The trials were not a great success for various reasons, and the service soon returned to Class 60/66 haulage. This is the same loco on the same train just a few days earlier, passing beneath the Poole High Street footbridge. The increase in rail traffic, and the number of times the crossing gates here are raised and lowered, has meant that it is used far more in recent times than it ever was when first placed here back in the mid-1870s! 20 December 2004.

This is a powerful view from a low perspective of the 7O51 Whatley Quarry to Hamworthy stone train passing through the centre road at Eastleigh station. 16 January 2008.

During a strange period of operating practices in the early 2010s, there was a brief time when one of the sub-class was used as yard shunter of all things at Eastleigh, the Class 08 shunters at the time were deemed not man enough to shunt the sometimes heavy loads around! This is the 3,300hp heavy freight loco pottering around with a few flat wagons in Eastleigh East Yard, which at the time was operated and maintained by DB Schenker, but it has since transferred to GBRf jurisdiction. 15 June 2012.

Vale of White Horse

Now in DB Schenker livery, 59202 is seen on a much more taxing and normal duty. This is 6M20, the 10.34 Whatley Quarry to St Pancras Churchyard Sidings, coming off the Berks & Hants line amongst the rooftops near Reading West. It is passing the old Reading Depot (directly behind the train), which has now been reduced to a stabling and maintenance point for track machines since the opening of the brand-new depot on the nearby Great Western Main Line in the mid-2010s. 16 December 2014.

Looking impressively clean in the cutting on the approach to Twyford on a fine autumn afternoon, the loco heads up 7C77, the 12.41 Acton to Merehead Quarry, as it makes its way west. Note again the re-tasked EWS coal hoppers making up the front half of the train. 1 October 2015.

Leaning to the cant through Newbury station on the up through line, this is 7A09, the 07.12 Merehead Quarry to Acton, heading east with yet another load from Somerset. This station has since been transformed with the coming of the overhead electrification. 19 May 2016.

With the DB Schenker red paintwork now starting to look a bit battered and worn, the loco is seen passing Styles Hill footbridge on the Frome avoiding line with 7A17, the 10.28 Merehead Quarry to Acton. This loco was admitted to Toton for its repaint into Freightliner orange and black livery a year after this image was taken. 12 August 2020.

Alan Meddows Taylor
MD, Mendip Rail Limited

Rounding the curve into Southampton Central, the loco has 66109 for company with 6O41, the 10.14 Westbury to Eastleigh East Yard infrastructure train. 66109 has since become a minor celebrity loco, when it was painted in a one off dark blue PD Ports colour scheme in 2019. 14 October 2011.

Nine years later, and this time, the loco has Freightliner's 66522 tucked inside as the crisp early morning sunshine nicely illuminates 7A09, the 07.12 Merehead to Acton again. The train is not far into its journey at Masters foot crossing near Fairwood Junction, Westbury. 26 August 2020.

Class 59/2

59203 *Vale of Pickering*

Seen through a 400mm lens on a dull day, this is 7A17, the 10.28 Merehead Quarry to Acton, approaching Taplow, an image now impossible to repeat today with the advent of the Great Western Main Line electrification. This loco was always recognisable at the time by the extra-large running number on the front end. 19 November 2008.

This is the 7O51 Whatley Quarry to Hamworthy loaded train passing the Freightliner terminal at Millbrook. This is another shot that is now impossible; the footbridge it was taken from, although still in situ, has been closed for safety reasons. 20 March 2009.

With the sunlight glinting off the side of the loco on a fine early autumn morning, this is 6M20, the 10.34 Whatley Quarry to St Pancras Churchyard Siding, passing the pumping house at Crofton on the Berks & Hants line. The Kennett and Avon Canal is in the foreground. 30 September 2011.

Recent trackside clearance hereabouts has afforded this long-lost view of the 7O51 Whatley Quarry to Hamworthy loaded train, just south of Ashurst in the New Forest. This location is more or less in the middle of nowhere, with only forest paths to allow access and a walk of perhaps 30 minutes or more from Ashurst station. Needless to say, the location has since become overgrown once again. 15 April 2009.

Now absolutely resplendent in DB Schenker's bright red livery and, unfortunately, now de-named, ex works 59203 passes West Drayton with 6M20, the 10.34 Whatley Quarry to St Pancras Churchyard Sidings. 31 August 2012.

Just over a year after the previous picture, the bright red DB Schenker livery is still very clean as the loco approaches Campbell Road bridge, Eastleigh, with 6O41, the 10.14 Westbury to Eastleigh East Yard. DB Schenker always seemed happy to use these locos on less demanding trains such as this one on occasion. 17 October 2013.

Despite the annoying overhead cables, this is an unusual shot of the loco passing through Westbury station light engine. This view has only been made possible by the recent demolition of the old British Rail Staff Association (BRSA) club buildings and the resultant extension to the car park. 7 March 2019.

A nice clear spring morning sees 7O12, the 03.12 Merehead Quarry to Woking loaded train, passing through Basingstoke. With a great deal of building work having taken place in the background, including large blocks of flats, this view looks rather different today. The image was taken from the top of a convenient multi-storey car park. 3 April 2017.

With the takeover of the whole class by Freightliner at the end of 2019, a start was made on repainting the Class 59/2s in the company's latest orange and black livery, although this stalled with the outbreak of the Coronavirus pandemic in early 2020. This was the second member of the class to receive the new livery, and is seen passing Heywood Village, soon after departing Westbury with 7B12, the 11.23 Merehead Quarry to Wootton Basset loaded train. 4 November 2020.

59203 is seen at Fairwood Junction again with 7A17, the 10.24 Merehead Quarry to Acton loaded train. This service usually goes via the avoiding lines on the left, but, on this occasion, there was a hot weather restriction in place on these lines, meaning it had to take the line through the station. Although initially dispensed with by DB Schenker, it is rather a shame that Freightliner has not retained the original names given to these locos. 30 June 2021.

The family likeness is evident between the Class 59 and the later built Class 66. We go back a few years and find 59203 has 66065 for company while working the lightly loaded 6O41, the 10.14 Westbury to Eastleigh East Yard, as they pass under the M3 motorway on the approach to Southampton Airport Parkway station, not far from its destination. 12 January 2012.

59204 *Vale of Glamorgan*

This is the 7O51 Whatley Quarry to Hamworthy stone train approaching Poole. Many years ago, the area under development to the right of the picture was the site of the huge Poole gasworks and was originally rail connected. It lay disused for many years until its complete demolition during the 1990s to make way for a major housing development. 7 January 2005.

59204 is seen approaching Fairwood Junction with 6M20, the 10.34 Whatley Quarry to St Pancras Churchyard Sidings. This train was always quite recognisable around this time, as it often employed this motley rake of ex-RMC hoppers. 25 September 2009.

The early bird gets the worm, as the saying goes! With a good coating of frost at sunrise on a very cold winter's morning at Westbury, this is the amazing sight of 59204 hauling 59104 *Village of Great Elm* and 59005 *Kenneth J Painter* on 7A09, the 07.12 Merehead Quarry to Acton. The train arrived here from Merehead with just 59104 and 59005, but the Class 59/2 was noted stabled at Westbury a short while earlier and was duly attached to the front. This is the only time I have ever seen all three Class 59 sub-classes on one train, which was almost certainly for loco balancing purposes. 20 January 2011.

The loco is now painted in DB Schenker's bright red livery with nameplates removed as it passes through Southampton Central with 7V07, the 13.41 Chichester to Merehead Quarry empties. 6 February 2018.

Oldfield Park in the Bath suburbs is the setting as 59204 heads 6A83, the 11.35 Avonmouth Bennetts Road to Westbury empty train, through the station. The buildings clearly show the type of stonework that the fine city of Bath is recognised for. 15 October 2019.

A fine autumn day sees the restarted (after just over a year's break) direct service from the Mendips to Chichester in Sussex. On the second day after the restart, the train is captured on its return, running as 7V07, the 11.41 Chichester to Merehead Quarry, and is seen crossing Bursledon Viaduct between Fareham and St Denys. Freight trains along this section of line are very few and far between these days, with this one being the only semi-regular service at the time of writing. 29 November 2012.

With the lock gates at Crofton prominent in the foreground on a fine spring morning, this is 59204 paired up with pioneer 59001 *Yeoman Endeavour* and heading west with 7C77, the 12.41 Acton to Merehead empty train. 8 April 2015.

59205 *Vale of Evesham* / L Keith McNair

Hauled by the penultimate loco of the class, this is the 7O51 Whatley Quarry to Hamworthy. Having just passed through Poole station, it is now approaching the former Holes Bay Junction, which used to see the line to Broadstone (where the Somerset and Dorset line branched off north to Bath), Ringwood and Brockenhurst, all long since closed, unfortunately. The train now has just a couple of miles to go until its destination is reached. 8 July 2004.

The fine autumn colours are starting to show as 59205 brings the 7B12 Merehead Quarry to Wootton Basset loaded train under the double span bridge at Fairwood Junction on the approach to Westbury. 29 October 2010.

Above: In a patch of weak sunshine, arriving at Westbury off the line from Trowbridge on a misty afternoon, this is 6A83, the 11.35 Avonmouth Bennetts Road to Westbury empty train. The lines to the right head off to Heywood Road Junction and the main line via Newbury up to London Paddington. 2 March 2012.

Right: A view now totally lost owing to the overhead catenary, this is 7A17, the 11.23 Merehead Quarry to Acton, on the approach to Reading West. The train has just passed Southcote Junction, where the Basingstoke to Reading line joins, and the Berks & Hants line to Newbury and beyond begins. 5 July 2012.

Could there be more of a contrast to a heavy stone train? This is the same 'Quantock Pullman' charter shown previously but a little earlier in the day passing Bapton, between Salisbury and Warminster. Strangely enough, this was not the first time that a member of the class had hauled this prestigious train, there were a few times when a '59/2' was used in and around the London area in the late 1990s and early 2000s, usually because of the unavailability of the rostered loco. 14 May 2016.

Another autumnal picture as we see 7O12, the 03.12 Merehead to Woking, passing through Fleet on the South Western Main Line, unusually on the fast line. This relatively recent view has been made possible by the construction of a multi-storey station car park next to the line in the mid-2010s. 11 November 2016.

Now this is what all railway embankments used to look like and indeed still should. Superb clearance work has been undertaken at Coker Woods on the Salisbury to Exeter line, just west of Yeovil Junction. However, the sight of a freight train here is even rarer than clear linesides! This is diverted 7V27, the Exeter Riverside to Westbury empty stone train, which was routed this way because the normal booked route via Taunton was blocked for engineering work in connection with Whiteball Tunnel. With the line singled way back in the late 1960s, it is now very hard to believe this was once a double tracked main line throughout. Regular freight trains ceased to run on this section of the former Southern Region West of England line many decades ago. 18 February 2019.

A very rustic view of 7A17, the 11.23 Merehead Quarry to Acton, approaching Lambert's Bridge near Westbury. 30 March 2021.

With 66053 along for the ride, a low autumn sun nicely highlights both the moody looking sky and 7C77, the 12.41 Acton to Merehead, waiting for a green signal at Westbury station. Nowadays, it is a bit of a lottery as to whether this train is routed either via the station or takes the avoider, and is often down to train crew requirements, as was the case here, waiting for a driver. 1 November 2018.

59206 *Pride of Ferrybridge / John F Yeoman*

Apparently designed to look like a railway signal box, the very unusual, almost fairground like, stone conveyor at Purley is the backdrop here as we see 59206 in the station after completion of unloading and shunting. It is preparing to depart with 6Y93, the 09.11 Purley to Cliffe Brett Marine empties. 28 January 2009.

The 7O51 Whatley Quarry to Hamworthy loaded train is seen approaching Dean on the Salisbury to Romsey line. Dean station and Mottisfont & Dunbridge, a few miles further down the line, were for a long time run by GWR but were served by SWR trains and no GWR services actually stopped there! One wonders how many stations in the UK had a similar claim to fame? The situation was finally put right in the late 2010s, with both these stations transferring to SWR ownership. 10 July 2008.

This loco was the first in the UK to be painted in the bright red DB Schenker livery in 2009 and is seen on one of its first duties in this new livery, passing St Denys with 7V07, the 11.41 Chichester to Westbury empty stone train. 25 February 2009.

The platform staff at Crowcombe Heathfield on the West Somerset Railway heritage line look a little bemused as the immaculate 3.300hp heavy freight loco awaits departure west to Minehead with a train from Bishops Lydeard during a diesel gala. 10 June 2011.

A far more common duty for the loco as we see 7A17, the 11.23 Merehead Quarry to Acton, taking the Westbury station avoiding line and clattering across the pointwork at Fairwood Junction. 25 September 2009.

This is 7V16, the 11.47 Fareham to Whatley Quarry, having just curved off the South Western Main Line at Eastleigh East Junction and taking the former freight only Romsey line westbound. 7 October 2010.

A super vista across part of the large expanse of Poole Harbour at Holes Bay can be had from the spur road flyover at Poole. This is empty 7V52, the 13.58 Hamworthy to Whatley. The wooded area above the train in the background is actually the small Pergins Island, which is uninhabited – at least by humans! 5 May 2011.

59206 heads south at Princes Risborough with 6A58, the 10.16 Calvert to Northolt empty binliner train. At this time, this was quite a regular turn for a '59/2' but has now been a Class 66 duty for some years. 10 November 2011.

Freight services on the section of line between Salisbury and Worting Junction are infrequent in the early 2020s, but one reliable working is 7O12, the 03.12 Merehead Quarry to Woking, which is seen approaching Grateley. 23 May 2013.

Above: Five months later and we see the same service as in the previous picture, but this time running as 6O12, the 03.23 Merehead Quarry to Woking. It was quite late at this point, passing Salisbury on a cracking autumn morning. 30 October 2013.

Right: A convenient footpath cuts across the rapeseed field at this location, as the loco comes under Lambert's Bridge on the approach to Westbury with 7A74, the 08.40 Merehead Quarry to Theale loaded train. 26 April 2018.

As mentioned previously, 59206 was the first of the class in DB Schenker livery in 2009, and, strangely enough, it was also the first Class 59 to receive the new Freightliner orange and black livery in early 2020. Soon after its makeover, it is seen at Masters foot crossing near Fairwood Junction with 7A77, the 12.03 Merehead Quarry to Theale. 26 August 2020.

Framed by the road bridge and making its way beneath a tangle of overhead wiring paraphernalia, this is 7C77, the 12.41 Acton to Merehead empties, approaching Twyford, just east of Reading on the Great Western Main Line. 7 April 2021.

Further reading from

As Europe's leading transport publisher, we produce a wide range of railway magazines and bookazines.

Visit: shop.keypublishing.com for more details